The Return to Dominate the Moon

A Journey through the First Commercial Lunar Landing, the Legacy of Odysseus, and the Dawn of Commercial Space Exploration

Mary E. Boswell

Table of Contents

Introduction: A Lunar Resurgence
The Long Absence of the United States from the Moon
Rise of Commercial Space Exploration

Chapter 1: Into the Unknown
The Quest for the Moon
Rise of Commercial Space Exploration

Chapter 2: Houston, We Have a Mission
Genesis of Intuitive Machines
Navigating the Challenges

Chapter 3: Technical Turbulence
Laser Malfunction and Mission-Endangering Hurdles
Ingenious Solutions from Earth to Lunar Orbit

Chapter 4: Odysseus Unveiled
The Odysseus Robot: Design and Technology
Preparing for the Descent

Chapter 5: Touchdown at the Southern Frontier
The Nerve-Wracking Wait for a Signal
Initial Data and Pictures from Lunar Soil

Chapter 6: The Power of Partnership
NASA's Collaboration with Intuitive Machines

A New Era of Public-Private Space Endeavors

Chapter 7: Craters, Shadows, and Frozen Water
Scientific Significance of the Lunar South Pole
Potential Resources for Future Human Exploration

Chapter 8: Payloads and Discoveries
NASA's Scientific Instruments on Odysseus
Commercial Contributions: Cameras and Art on the Moon

Chapter 9: The Lunar Dust Dilemma
Investigating the Behavior of Lunar Dust
Lessons from Apollo and Plans for the Future

Chapter 10: Beyond the Moon
Implications for Future Space Exploration
Commercial Ventures in the Cosmos

Chapter 11: A Lunar Legacy
Reflecting on the Historic Achievement
Inspirations for the Next Generation

Conclusion: The Odyssey Continues
Looking Ahead to the Next Frontier
The Ever-Expanding Universe of Commercial Space Exploration

Introduction: A Lunar Resurgence

The cold void of space has long held humanity's fascination, and no celestial body has captivated our imaginations quite like the Moon. For decades, it stood as both a symbol of triumph and a testament to the heights of human achievement. Yet, as the years passed, the lunar surface remained untouched by American hardware, leaving a void in the nation's space legacy.

The Long Absence of the United States from the Moon

The Moon, Earth's celestial companion, witnessed a flurry of human activity during the Apollo era. Between 1969 and 1972, a series of manned missions, collectively known as Apollo, marked the zenith of American space exploration. Astronauts set foot on the lunar surface, planting the stars and stripes, conducting experiments, and etching the indelible footprint of humanity on the desolate terrain.

However, after the final Apollo mission in 1972, the lunar landscape fell silent, devoid of American presence for over half a century. The reasons behind this prolonged absence were multifaceted, ranging from shifting priorities in space exploration to budget constraints and geopolitical changes. As other nations made strides in lunar exploration, the United States redirected its focus toward other realms of the cosmos, such as the Space Shuttle program and the exploration of Mars.

This prolonged hiatus, while punctuated by robotic missions and scientific endeavors, left a lingering question: when would the United States return to the Moon? The answer to this question would not only shape the trajectory of American space exploration but also herald a new era of lunar exploration driven by a dynamic force – the rise of commercial space exploration.

Rise of Commercial Space Exploration

As the 21st century unfolded, a paradigm shift in space exploration dynamics took center stage. No longer confined solely to the realms of governmental agencies, the exploration of space opened its doors to private enterprises. The emergence of visionary entrepreneurs like Elon Musk with SpaceX and Jeff Bezos with Blue

Origin signaled a departure from traditional models of space exploration.

In this new landscape, companies sought not only to deliver payloads to space but to push the boundaries of what was deemed achievable. The Moon, with its potential for scientific discovery and resource utilization, became an enticing destination. The seeds of commercial lunar exploration were sown, and companies began to vie for a coveted spot in the race to reach Earth's natural satellite.

Houston-based Intuitive Machines emerged as a key player in this lunar resurgence. Armed with a vision to break the shackles of governmental exclusivity, Intuitive Machines embarked on a journey that would redefine the possibilities of commercial space exploration. Their mission was ambitious: to not only land a spacecraft on the Moon but to do so with a robot capable of conducting meaningful scientific experiments.

The rise of commercial space exploration brought with it a sense of democratization in the cosmos. No longer limited by the constraints of government budgets and bureaucratic timelines, private companies infused a spirit of innovation and efficiency into space missions. The competitive spirit among these entities propelled advancements in rocket technology, spacecraft design, and mission planning, setting the stage for Intuitive Machines' historic lunar touchdown.

Chapter 1: Into the Unknown

The Quest for the Moon

The Moon, Earth's constant companion, has been a celestial muse for explorers and dreamers throughout history. From ancient myths to scientific curiosity, the Moon has played a pivotal role in shaping human perceptions of the cosmos. However, it was in the mid-20th century that the dream of reaching the Moon transformed into a tangible goal.

The quest for the Moon reached its zenith with the Apollo program. Conceived in the 1960s, against the backdrop of the Cold War space race, the Apollo missions represented a monumental leap in human exploration. The dream of setting foot on the lunar surface became a reality when Apollo 11, with astronauts Neil Armstrong and Buzz Aldrin, achieved the historic feat on July 20, 1969. Subsequent Apollo missions continued to unravel the mysteries of the Moon, leaving an indelible mark on the pages of human history.

However, as the Apollo program concluded in 1972, the lunar landscape fell silent. The footprints of astronauts remained the only human imprints on the desolate surface for decades. The quest for the Moon, though never entirely abandoned, entered a phase of dormancy as other frontiers of space captured the attention of space agencies.

Rise of Commercial Space Exploration

Amidst the quiet lunar hiatus, a paradigm shift was underway in the world of space exploration. The 21st century witnessed the emergence of a new breed of space pioneers - private companies fueled by visionaries determined to democratize space. This marked the rise of commercial space exploration, a movement that would reshape the dynamics of cosmic exploration.

Pioneers like Elon Musk, with SpaceX, and Jeff Bezos, with Blue Origin, became the vanguard of this new era. Their ambitious goals extended beyond Earth's atmosphere, aiming for destinations that were once solely within the purview of governmental space agencies. The Moon, with its proximity and potential for scientific discoveries and resource utilization, emerged as a tantalizing target for these commercial ventures.

Enter Houston-based Intuitive Machines, a company poised to break the shackles of traditional space exploration models. Their vision was clear: to become the first commercial entity to land a spacecraft on the Moon. The quest for the Moon, dormant in the national consciousness, was rekindled by the audacity and ambition of private enterprises.

The competitive spirit among these companies mirrored the fervor of the historic space race, but this time, the finish line was not limited to national pride. It was about unlocking the economic and scientific potential of our celestial neighbor. As Intuitive Machines geared up for its lunar mission, the echoes of a new space era reverberated through the corridors of exploration, signaling a departure from the conventional and embracing the unknown possibilities of commercial space ventures.

Chapter 2: Houston, We Have a Mission

Genesis of Intuitive Machines

The genesis of Intuitive Machines lies in the convergence of technological innovation, entrepreneurial spirit, and a steadfast commitment to push the boundaries of space exploration. Founded in 2013 by Steve Altemus, a veteran in the aerospace industry, the company was born out of a desire to bridge the gap between commercial enterprise and lunar exploration. Houston, a city renowned for its association with NASA's Johnson Space Center, became the fertile ground where the seeds of Intuitive Machines took root.

The company's mission was audacious: to become the first commercial entity to land a spacecraft on the Moon. This goal, however, was not solely about achieving a historic milestone; it was a testament to the changing landscape of space exploration. The genesis of Intuitive Machines marked a departure from traditional governmental exclusivity in space endeavors, introducing a new era where private enterprises would play a pivotal role in shaping the cosmic narrative.

As Intuitive Machines took its first steps, the team faced the daunting challenge of not only designing and building a spacecraft capable of lunar descent but also securing the necessary partnerships and resources to turn their vision into a reality. The genesis of Intuitive Machines encapsulates the spirit of innovation and determination that defines the ethos of the company.

Navigating the Challenges

Space exploration is fraught with challenges, and for Intuitive Machines, navigating these challenges became an integral part of their mission. Even before the Odysseus robot embarked on its lunar journey, the team encountered a technical hurdle that threatened to halt the mission before it could commence.

The ranging lasers, designed to calculate the spacecraft's altitude and velocity during descent, faced malfunctions. These lasers were critical for ensuring a controlled and precise landing on the lunar surface. The initial setback led to a moment of reckoning for the team at Intuitive Machines. The success of their mission hung in the balance, and the resilience of their vision would be tested in the crucible of space.

Faced with this challenge, the team displayed a remarkable blend of resourcefulness and adaptability. Experimental lasers from NASA, conveniently on board the spacecraft, emerged as unexpected saviors. Engineers ingeniously patched these lasers into the navigation computers, overcoming the malfunction and allowing the mission to proceed. The ability to navigate and surmount challenges is an inherent aspect of any space exploration endeavor, and Intuitive Machines showcased their prowess in adapting to the unpredictable nature of cosmic exploration.

The navigation through challenges extended beyond technical glitches. The vastness of space presented its own set of obstacles, including the nerve-wracking minutes of waiting for a signal to confirm Odysseus' successful lunar touchdown. The delicate dance between the robotic explorer and the lunar surface unfolded against the backdrop of the unknown, emphasizing the unpredictable nature of space missions.

Chapter 3: Technical Turbulence

Laser Malfunction and Mission-Endangering Hurdles

The journey to the Moon is a precarious dance between meticulous planning and the unpredictable nature of space. As the Odysseus spacecraft descended toward the lunar surface, the team at Intuitive Machines encountered a daunting challenge – the malfunction of the ranging lasers. These lasers, designed to calculate altitude and velocity during descent, were the linchpin for a safe and controlled landing.

In the face of this unexpected technical turbulence, the mission hung in the balance. The success of becoming the first commercial entity to land a spacecraft on the Moon teetered on the edge of uncertainty. The potential mission-ending hurdle cast a shadow over the Houston-based team, echoing the historical challenges faced by lunar explorers in the past.

The stakes were high. Failure to resolve the laser malfunction would not only jeopardize the mission's success but could also have broader implications for the credibility of commercial space exploration. It was a pivotal moment for Intuitive Machines, a test of their mettle and the resilience of their vision.

Ingenious Solutions from Earth to Lunar Orbit

Amid the shadows of uncertainty, the team at Intuitive Machines showcased their ingenuity, turning what seemed like a mission-ending obstacle into an opportunity for innovation. Onboard the spacecraft were experimental lasers from NASA, an unforeseen contingency that proved to be a lifeline for the mission.

The engineers, working tirelessly from Earth, devised an ingenious solution. They patched the experimental lasers into the navigation computers of the Odysseus robot, creating a makeshift yet effective alternative to the malfunctioning lasers. This on-the-fly adaptation in the face of technical adversity not only demonstrated the technical prowess of the team but also highlighted the collaborative nature of space exploration.

The journey from Earth to lunar orbit became a saga of problem-solving that transcended the vastness of space. Ingenious solutions echoed the spirit of the Apollo era, where quick thinking and resourcefulness were the hallmarks of successful space missions. The ability to adapt, innovate, and overcome technical hurdles showcased the evolving nature of space exploration, now led not only by governments but also by nimble and innovative commercial entities.

The technical turbulence, which could have been a mission-ending storm, transformed into a testament to human ingenuity. As the Odysseus spacecraft continued its descent, the makeshift solution held, and the lunar surface approached with cautious optimism. The chapter not only unravels the technical intricacies of the laser malfunction but serves as a pivotal moment in the broader narrative of commercial space exploration.

Chapter 4: Odysseus Unveiled

The Odysseus Robot: Design and Technology

At the heart of Intuitive Machines' lunar mission stood the Odysseus robot, a testament to human ingenuity and technological innovation. The chapter begins by peeling back the layers of the robotic explorer, delving into its design and the cutting-edge technology that made it a pioneering force in lunar exploration.

The design of Odysseus was a meticulous fusion of functionality and resilience. Shaped by the harsh lunar environment and the demands of a controlled descent, the robot boasted a sturdy frame equipped to withstand the rigors of landing on an alien surface. Its mobility, crucial for conducting scientific experiments and navigating the challenging lunar terrain, was a key consideration in the design process.

The technological prowess of Odysseus extended beyond its physical structure. Equipped with state-of-the-art sensors, cameras, and scientific instruments, the robot was designed to gather crucial data about the lunar environment. The significance of this data was not only scientific but also laid the groundwork for future human exploration, providing insights into the conditions that astronauts might encounter.

As Odysseus was unveiled to the world, it stood as a symbol of the evolving nature of space exploration. No longer confined to the realms of governmental agencies, advanced robotics like Odysseus marked a new era where commercial entities played a pivotal role in shaping the future of cosmic exploration.

Preparing for the Descent

The descent to the lunar surface was the culmination of years of planning, engineering, and anticipation. As the Odysseus robot stood poised on the brink of making history, the chapter details the meticulous preparations undertaken by the team at Intuitive Machines to ensure a successful landing.

Every detail, from the trajectory of the descent to the timing of critical maneuvers, was scrutinized with precision. The engineers and mission control teams worked in tandem, their focus unwavering as they fine-tuned the parameters that would guide Odysseus through the final leg of its cosmic journey.

Simulation after simulation was conducted, recreating the lunar descent in various scenarios. These simulations served not only as technical rehearsals but also as a means to anticipate and plan for any unforeseen challenges that might arise. The preparations for the descent were not just about reaching the Moon; they were about ensuring the safe arrival and functionality of the Odysseus robot on the lunar surface.

The chapter unfolds the tension and excitement within mission control as the clock ticks down to the critical moment. The team's dedication and expertise were on full display as they navigated the complexities of lunar descent, a feat not attempted by an American entity in over five decades.

Chapter 5: Touchdown at the Southern Frontier

The Nerve-Wracking Wait for a Signal

With bated breath and hearts echoing the pulse of anticipation, mission control at Intuitive Machines entered a phase of nerve-wracking suspense. The descent of Odysseus towards the lunar surface marked the culmination of meticulous planning, technical innovation, and the audacious vision to rekindle American presence on the Moon.

As the clock ticked down, every second felt like an eternity. The vast expanse of space introduced inherent communication delays, leaving mission control in an agonizing wait for the crucial signal indicating a successful lunar touchdown. The historical significance of this moment was palpable, transcending the technicalities and echoing the echoes of Apollo-era glory.

The silence in the control room spoke volumes as engineers and scientists exchanged glances, each second stretching the bounds of anticipation. The unknowns of space, coupled with the inherent complexities of lunar descent, amplified the tension. The wait for a signal became a testament to the unpredictability and high stakes inherent in cosmic exploration.

The nerves of the team at Intuitive Machines were not unfounded. The lunar environment is unforgiving, and the historical absence of American hardware on the Moon for over fifty years added an extra layer of weight to the moment. The echoes of the Apollo missions lingered in the background, making this lunar touchdown not just a technological feat but a symbolic return to a frontier that had been untouched for far too long.

Initial Data and Pictures from Lunar Soil

Then, the moment of truth arrived. A faint signal, a blip on the screens, heralded the triumph of Odysseus. The robotic explorer had successfully landed on the lunar surface, marking the first commercial entity to achieve this milestone. The collective sigh of relief in mission control was drowned by cheers and applause, echoing through the Houston-based headquarters.

As the initial data started streaming in, the screens lit up with a cascade of information from Odysseus. Telemetry data, sensor readings, and system statuses confirmed not only the success of the landing but also the functionality of the robotic explorer on the lunar soil. The nerves that had gripped the control room transformed into jubilation, a celebration of a mission accomplished against the backdrop of the lunar unknown.

Pictures from the lunar surface followed, providing a visual narrative of the historic landing. The landscape, stark and barren, unfolded before the eyes of the team, capturing the essence of an untouched celestial frontier. The images from Odysseus' perspective became a window into a world that had been absent from the human gaze for decades.

The targeted landing site, nestled near the southern pole of the Moon, showcased a cratered terrain next to a towering mountain complex known as Malapert. This region, unexplored by American hardware since the Apollo era, became the stage for a new chapter in lunar exploration. The success of the mission added not only scientific value but also opened up possibilities for future human exploration.

The southern pole of the Moon holds unique significance. Its craters, some perpetually shrouded in darkness, are believed to contain frozen water. The potential presence of water on the Moon has profound implications for future lunar missions, providing a valuable resource for sustenance and fuel. The success of Odysseus at this southern frontier laid the groundwork for understanding and harnessing these resources for future space exploration endeavors.

Chapter 6: The Power of Partnership

NASA's Collaboration with Intuitive Machines

The collaboration between NASA and Intuitive Machines became the linchpin of the Odysseus mission. While Intuitive Machines spearheaded the endeavor with a vision to break the commercial barrier on lunar exploration, NASA brought its wealth of experience, scientific expertise, and a legacy of space exploration to the table.

NASA's involvement was not merely a financial backing; it was a strategic partnership that leveraged the strengths of both entities. The U.S. space agency purchased room on the Odysseus spacecraft for six scientific instruments, a payload that represented the collective ambitions of NASA's scientific community. This partnership symbolized a shared vision: to push the boundaries of lunar exploration and pave the way for future human endeavors.

The collaboration extended beyond financial support. NASA's technical guidance, experience in deep space exploration, and intricate knowledge of the lunar environment were invaluable assets to Intuitive Machines. The expertise that NASA brought to the table served as a guiding beacon, especially when technical challenges threatened to jeopardize the mission.

Intuitive Machines, as a commercial entity, epitomized the agile, innovative spirit of the private sector. NASA, with its decades-long legacy, provided a stabilizing force, ensuring that the mission adhered to the highest scientific standards. The collaboration showcased the power of merging the dynamism of the commercial sector with the seasoned wisdom of a government space agency.

A New Era of Public-Private Space Endeavors

The collaboration between NASA and Intuitive Machines marked a paradigm shift in the landscape of space exploration. No longer confined to the realm of governmental exclusivity, space endeavors were evolving into collaborative ventures, where the strengths of public and private entities harmonized for the greater good of cosmic exploration.

The Odysseus mission exemplified the potential of such collaborations. It demonstrated that public-private partnerships could not only break new ground in exploration but also rekindle the flames of ambition and possibility. The collaboration embodied a departure from the traditional model where space exploration was solely the purview of government agencies.

As NASA's administrator, Bill Nelson, remarked, "Today is the day that shows the power and promise of NASA's commercial partnerships." The successful landing of Odysseus was not just a triumph for Intuitive Machines but a validation of the efficacy of collaborative space endeavors. It echoed a sentiment that had been gaining momentum in the space community—that the synergy between governmental agencies and commercial entities could amplify the impact of space exploration.

This collaborative model is not limited to a single mission. It paves the way for a future where the frontiers of space are explored through a fusion of governmental resources, expertise, and the nimbleness of commercial innovation. The Odysseus mission became a torchbearer for this new era, illuminating a path where collective endeavors transcend national boundaries and corporate interests, uniting humanity in the quest for knowledge beyond our planet.

Chapter 7: Craters, Shadows, and Frozen Water

Scientific Significance of the Lunar South Pole

The southern pole of the Moon stands as a celestial crucible of scientific curiosity and potential exploration. Its unique characteristics and relative isolation from direct sunlight make it a region of exceptional scientific interest. As the Odysseus robot touched down in this uncharted territory, it brought with it the promise of unlocking secrets hidden in the craters and shadows of this lunar frontier.

The craters near the southern pole of the Moon are particularly intriguing. Some of these craters, owing to the Moon's axial tilt, remain in perpetual shadow. These permanently shadowed regions (PSRs) have captivated scientists for years because they offer a unique environment where temperatures are extraordinarily low, reaching as low as minus 250 degrees Fahrenheit (minus 157 degrees Celsius). This extreme cold creates conditions that are believed to preserve volatile compounds, including water ice, over geological timescales.

Scientists theorize that these PSRs could harbor significant amounts of frozen water. The presence of water on the Moon has profound implications for future lunar exploration. Water is not only a critical resource for sustaining human life, but it can also be converted into oxygen and hydrogen, essential components for life support systems and rocket fuel. The southern pole, with its perpetually shadowed craters, becomes a treasure trove of potential resources that could support sustained human presence on the Moon.

The scientific significance of exploring the lunar south pole extends beyond resource utilization. It serves as a time capsule, preserving a record of volatile compounds that could offer insights into the early history of our solar system. By studying these shadowed regions, scientists aim to unravel the mysteries of lunar formation, evolution, and the processes that shaped our celestial neighbor over billions of years.

Potential Resources for Future Human Exploration

As the Odysseus robot stands on the lunar soil near the southern pole, it heralds the prospect of utilizing the potential resources hidden in the shadows. The presence of frozen water is not just a scientific curiosity; it is a strategic asset for future human exploration.

Water is essential for sustaining life, and the ability to harness lunar water resources would alleviate the need to transport large quantities of water from Earth for future lunar missions. This reduction in payload could significantly lower the costs and logistical challenges associated with lunar exploration. Furthermore, the conversion of water into oxygen and hydrogen could provide a locally sourced supply for breathing air and rocket fuel, reducing dependency on Earth for essential resources.

The southern pole becomes a promising site for establishing a sustainable human presence on the Moon. Its unique combination of scientific intrigue and potential resource abundance positions it as a prime candidate for future lunar habitats. Establishing a foothold in this region could serve as a stepping stone for further exploration of the Moon and beyond, unlocking the door to a new era of human space exploration.

Chapter 8: Payloads and Discoveries

NASA's Scientific Instruments on Odysseus

NASA's collaboration with Intuitive Machines went beyond providing financial support; it extended to embedding a suite of six scientific instruments aboard the Odysseus spacecraft. Each instrument represented a scientific discipline, aiming to unravel specific mysteries of the Moon's composition, environment, and geological history.

One of the key investigations involved studying the behavior of lunar dust. The Apollo astronauts encountered challenges with lunar dust, which proved abrasive and had a tendency to cling to equipment. The Odysseus mission sought to deepen our understanding of this phenomenon, observing how lunar dust is kicked up during the descent, lingers above the surface, and eventually settles back down. Such insights are crucial for planning future lunar missions, especially those involving human exploration, where the impact of lunar dust on equipment and habitats is a significant consideration.

Another instrument focused on the lunar surface's thermal properties, shedding light on temperature variations across different terrains. Understanding these variations is vital for planning the deployment and operation of future instruments and equipment on the Moon.

NASA's payloads on Odysseus also included instruments designed to measure the Moon's exosphere, the tenuous outermost layer of its atmosphere. By analyzing the composition and density of the exosphere, scientists aimed to gain insights into the Moon's history and its interactions with the solar wind.

The collaboration between Intuitive Machines and NASA, manifested through these scientific instruments, elevated the Odysseus mission beyond a mere lunar touchdown. It transformed the robotic explorer into a mobile scientific laboratory, unraveling the mysteries of the Moon and contributing valuable data for future lunar endeavors.

Commercial Contributions: Cameras and Art on the Moon

While NASA's scientific instruments were the backbone of Odysseus' mission, commercial contributions added a layer of innovation and human expression to the lunar landscape. Cameras, both scientific and artistic, were among the commercial payloads that accompanied Odysseus on its journey.

Embry-Riddle Aeronautical University contributed a student camera system designed to capture the robot's descent and landing. The camera system, intended to operate when Odysseus was still 30 meters above the lunar surface, aimed to provide a unique perspective, capturing selfie images as the robot set itself down. This commercial payload not only served a scientific purpose but also brought a touch of youthful exuberance to the mission, involving the next generation of space enthusiasts in the historic lunar endeavor.

In a move that fused technology and art, American artist Jeff Koons attached a box to the side of the lander. This box contained 125 small stainless steel balls, each representing the Moon's different phases throughout a month. This artistic payload added a symbolic and aesthetic dimension to the mission, marking the Moon not just as a scientific canvas but also as a space for creative expression.

The inclusion of commercial contributions underscored the evolving landscape of space exploration. It demonstrated that the Moon is not only a realm of scientific inquiry but also a canvas for human creativity. The cameras and art on the Moon became a testament to the expanding horizons of lunar exploration, where diverse perspectives and contributions converge to enrich our understanding of the cosmos.

Chapter 9: The Lunar Dust Dilemma

Investigating the Behavior of Lunar Dust

As the Odysseus robot descended towards the lunar surface, it carried with it a scientific mission that sought to shed light on the elusive and complex behavior of lunar dust. The challenge posed by lunar dust was not new; it echoed the experiences of Apollo astronauts who grappled with its abrasive nature and propensity to cling to equipment.

The investigation into lunar dust behavior aimed to address critical questions. How does lunar dust behave when disturbed during the descent of a spacecraft? What are the mechanisms that cause it to linger above the lunar surface before settling back down? These questions were not merely academic; they held practical implications for future lunar missions, particularly those involving human exploration.

The behavior of lunar dust is intricately tied to the Moon's unique environment. With no atmosphere to cushion its fall, lunar dust behaves differently than terrestrial dust. It can be electrostatically charged, creating an affinity to cling to surfaces. Understanding these nuances is crucial for designing equipment, spacesuits, and habitats that can withstand the lunar environment without being compromised by the abrasive and adhesive nature of lunar dust.

The Odysseus mission, equipped with scientific instruments dedicated to studying lunar dust, became a pivotal chapter in this ongoing quest for understanding. By capturing data during the descent and landing, the mission aimed to unravel the dynamics of lunar dust particles in the lunar exosphere. The findings held the potential to inform future engineering decisions and enhance the durability and effectiveness of equipment deployed on the lunar surface.

Lessons from Apollo and Plans for the Future

The challenges posed by lunar dust were not unknown to the scientific community, thanks to the valuable lessons learned from the Apollo missions. The experiences of astronauts, coupled with data collected from lunar modules and surface experiments, offered a foundation of knowledge. However, the lunar dust dilemma remained an open question, and the Odysseus mission sought to deepen our understanding.

One of the critical lessons from Apollo was the need for innovative engineering solutions to mitigate the impact of lunar dust. Seals, joints, and other mechanical components had to be carefully designed to withstand the abrasive nature of the lunar regolith. The Odysseus mission provided an opportunity to build upon this knowledge and explore new strategies for coping with lunar dust challenges.

The plans for the future are imbued with the spirit of adaptation and innovation. The Odysseus mission represents a stepping stone, and the insights gained pave the way for more sophisticated robotic and human missions to the Moon. As humanity sets its sights on returning to our celestial neighbor, the lunar dust dilemma becomes a challenge to be met with ingenuity and foresight, a puzzle to be solved for the advancement of lunar exploration.

Chapter 10: Beyond the Moon

Implications for Future Space Exploration

The success of the Odysseus mission holds profound implications for the trajectory of future space exploration. Beyond the immediate triumph of becoming the first commercial entity to land on the Moon, the mission serves as a catalyst for broader ambitions and possibilities in our quest to understand and traverse the cosmos.

First and foremost, the Odysseus mission contributes invaluable data to our understanding of the Moon. The scientific instruments on board provide insights into lunar dust behavior, thermal properties, and exospheric conditions. This knowledge becomes a foundation for planning and executing future lunar missions, especially those involving human exploration and habitation.

The southern pole, with its shadowed craters potentially harboring frozen water, emerges as a strategic location for future lunar habitats. The success of Odysseus opens the door to sustained human presence on the Moon, where resources like water can be harnessed to support life and enable further exploration of the lunar surface.

The data collected by Odysseus also holds relevance for upcoming robotic and crewed missions to Mars. The lessons learned from lunar exploration, including technological adaptations for dealing with lunar dust and extreme temperature variations, inform the engineering and planning of missions destined for the Red Planet.

Beyond the practical implications, the Odysseus mission ignites a renewed spirit of exploration. It serves as a testament to human ingenuity and determination, showcasing the collaborative efforts between governmental and commercial entities in pushing the boundaries of cosmic discovery. The success of Odysseus resonates as a beacon, beckoning humanity to venture farther into the cosmos.

Commercial Ventures in the Cosmos

The Odysseus mission marks a milestone in the evolution of commercial ventures in space. As the first commercial entity to land on the Moon, Intuitive Machines exemplifies the expanding role of private companies in space exploration.

This paradigm shift opens up new avenues for collaboration, innovation, and economic opportunities beyond Earth's atmosphere.

The collaboration between NASA and Intuitive Machines, as showcased in the Odysseus mission, sets a precedent for future public-private partnerships. It demonstrates that the combined strengths of governmental expertise and commercial agility can unlock new frontiers in space exploration. The success of this collaboration encourages further collaborations, not just for lunar exploration but for endeavors that extend to the far reaches of the solar system and beyond.

Commercial ventures in the cosmos are not confined to governmental collaborations alone. The burgeoning space industry involves a spectrum of activities, from satellite launches to asteroid mining, space tourism, and the development of space habitats. Companies are vying to establish a presence in low Earth orbit, with ambitions to provide services ranging from telecommunications to Earth observation.

The lunar success of companies like Intuitive Machines adds momentum to the commercial space sector. It fuels a competitive and innovative environment where companies strive to pioneer new technologies, reduce launch costs, and offer unique services. This surge in commercial activity in space has the potential to democratize access to space, making it more accessible to a broader range of entities and nations. The Moon becomes not just a destination but a stepping stone, a proving ground for technologies and business models that will be integral to future space endeavors.

The Odysseus mission, with its historic landing and scientific achievements, is a chapter in the larger narrative of humanity's cosmic journey. It beckons us to look beyond the Moon, to envision a future where the frontiers of space are explored not only by nations but by a diverse array of commercial entities.

Chapter 11: A Lunar Legacy

Reflecting on the Historic Achievement

As the echoes of cheers and applause reverberate through mission control, the Odysseus mission etches its mark on the lunar landscape and the collective memory of humanity. The robotic explorer, standing proud on the southern pole of the Moon, becomes more than a scientific instrument; it becomes a symbol of human ingenuity, collaboration, and the relentless pursuit of the unknown.

Reflecting on this historic achievement involves a journey through the challenges overcome, the uncertainties navigated, and the triumph of the human spirit in the face of cosmic frontiers. The Odysseus mission breaks a half-century silence, marking the return of American hardware to the lunar surface and the dawn of a new era in space exploration.

The success of the Odysseus mission resonates as a testament to the power of collaboration between governmental and commercial entities. It exemplifies a model where NASA's legacy and expertise join forces with the innovation and agility of private companies to push the boundaries of exploration. The southern pole of the Moon, once untouched since the Apollo era, now stands as a testament to the continual evolution of human space exploration.

The reflective journey extends beyond the technical achievements to the significance of returning to the Moon. The Moon, our celestial companion, has been a source of fascination, inspiration, and a symbol of human potential throughout history. The Odysseus mission reignites the flames of lunar exploration, not just as a scientific endeavor but as a beacon guiding humanity toward future cosmic endeavors.

Inspirations for the Next Generation

As readers traverse the pages of this chapter, they encounter not just a historical narrative but a call to action for the next generation of space enthusiasts. The Odysseus mission, with its triumphs and challenges, becomes a source of inspiration for those who will inherit the cosmic legacy.

The robotic explorer's footprint on the lunar soil becomes a metaphorical baton passed from one generation to the next. It carries with it the dreams of those who dared to reach for the stars, encouraging the next wave of explorers to envision even greater heights. The lunar legacy of Odysseus extends beyond the robotic explorer; it extends to the minds and aspirations of the students, scientists, engineers, and dreamers who contributed to and witnessed this historic moment.

Inspirations for the next generation are not confined to the technical prowess demonstrated by the Odysseus mission. It extends to the ethos of exploration—the courage to venture into the unknown, the resilience to overcome obstacles, and the curiosity that fuels the human spirit. The southern pole of the Moon becomes a rallying point for the imagination, inviting the next generation to dream big, think boldly, and reach for the stars.

The Odysseus mission, with its commercial collaborations and scientific achievements, presents a model for the future. It showcases that the cosmos is not an exclusive domain of governmental agencies but a frontier open to the participation of diverse entities, from private companies to educational institutions. The next generation is poised to inherit a cosmos ripe with possibilities, where lunar exploration is a stepping stone to even grander cosmic adventures.

Conclusion: The Odyssey Continues

Looking Ahead to the Next Frontier

The odyssey of lunar exploration, sparked by the historic achievement of the Odysseus mission, extends its gaze toward the next frontier. The Moon, once a distant and desolate neighbor, now stands as a waypoint in humanity's journey into the cosmos. The lessons learned, the data collected, and the collaborative spirit demonstrated during the mission become the building blocks for future endeavors.

Looking ahead involves envisioning a future where humans not only return to the Moon but establish a sustained presence. The southern pole, with its shadowed craters potentially harboring frozen water, becomes a strategic outpost for scientific exploration, resource utilization, and a launchpad for missions bound for even more distant destinations.

The Artemis program, spearheaded by NASA, aims to return humans to the lunar surface and, for the first time, include the first woman and the next man. This ambitious undertaking builds upon the legacy of the Odysseus mission, emphasizing the importance of lunar exploration as a stepping stone for future crewed missions to Mars and beyond.

The Ever-Expanding Universe of Commercial Space Exploration

The conclusion of the lunar odyssey is not the end but a prelude to the ever-expanding universe of commercial space exploration. The

Odysseus mission, as the first commercial entity to land on the Moon, marks a paradigm shift in the space industry. It exemplifies the potential for collaboration between governmental agencies and private companies, showcasing the role of commercial entities in advancing our exploration of the cosmos.

The success of companies like Intuitive Machines paves the way for an era where commercial ventures extend humanity's reach into space. From satellite launches and space tourism to asteroid mining and the development of space habitats, the commercial space sector becomes a dynamic force driving innovation, reducing launch costs, and democratizing access to space.

The lunar legacy of Odysseus becomes a beacon for commercial entities, inspiring them to venture beyond Earth's orbit and participate in the broader narrative of space exploration. The Moon, with its proximity and strategic importance, becomes a testing ground for technologies and business models that will shape the future of commercial space endeavors. The Moon, with its ancient mysteries and potential resources, becomes a focal point in our quest to understand the universe. Commercial entities, driven by ambition and innovation, emerge as integral players in this cosmic saga, transcending the boundaries of our home planet and reaching for the stars.

As the Odysseus mission takes its place in the archives of lunar exploration, it leaves behind not just footprints on the lunar soil but a legacy of collaboration, discovery, and the unyielding human spirit. The odyssey continues, fueled by the dreams of those who dare to explore the cosmos, and the next chapters are poised to unfold in the ever-expanding universe of commercial space exploration. The conclusion becomes a preface to the unwritten cosmic adventures that beckon humanity to reach new heights, explore new frontiers, and continue the timeless journey into the great expanse of the universe.